The Roots of Dentistry

The Roots of Dentistry

Edited by

Christine Hillam

for

The Lindsay Society for the History of Dentistry

1990

Published by the British Dental Association
64 Wimpole Street, London W1M 8AL

ISBN 0 904588 25 4

Typeset by Latimer Trend & Company Ltd, Plymouth
Printed in Great Britain by William Clowes Ltd, Beccles

Contributors

Margaret A. Clennett, BA, ALA
E. Muriel Cohen, *MBE*, BA
Ronald A. Cohen, MA, FFD
J. A. Donaldson, BA, FDS
Stanley Gelbier, PhD, LDS, DHMSA
A. S. Hargreaves, PhD, BDS
Christine Hillam, PhD, BA
D. G. Hillam, MDS, FDS
J. E. McAuley, LDS
David Wright, MA

Contents

Acknowledgements   viii

Preface   ix

1   Dental Disease   1

2   Dental Treatment   5

3   The Providers of Dental Treatment   35

4   The Patients   47

5   Dental Literature   55

Appendix 1   References cited in the text   62

Appendix 2   Chronology   63

Appendix 3   Glossary   67

Index   70

Acknowledgements

Dr Hillam and the Lindsay Society would like to thank David Wright, formerly the Assistant Keeper of the Wellcome Museum for the History of Medicine at the Science Museum, London, for providing all the photographs. The source of the specimens, equipment, bookplates, titles and cartoons are given in the individual captions.

Preface

A number of authoritative works have been produced since the beginning of the century, dealing with the history and development of dentistry. Unfortunately, most are not readily available outside specialist libraries.

This book has been compiled by members of the Lindsay Society (the historical society of the British Dental Association) to provide an introductory outline for general readers whose interest may perhaps have been aroused by museum exhibits, for dentists interested in the development of their own profession and for the growing number of schoolchildren seeking material for school projects.

In no way is it intended to replace any of the fuller expositions of the subject, since it must of necessity be selective. It is hoped that it will provide a basic framework of information for those who may wish to follow up some aspects further.

A glossary of technical terms is to be found on pages 67–69.

1

Dental Disease

Dental disease is as old as mankind. It has been found in the teeth of human remains from as long as 500 000 years ago. While it is largely confined to humans, it also occurs in domestic animals.

Early people lived on coarse foods, often contaminated with a liberal helping of grit. As a result, their teeth became very worn down, even to the extent that the sensitive pulps were exposed (fig. 1.1). Decay or, more scientifically, dental caries, was comparatively rare, but skulls frequently show a great deal of loss of bone around the teeth resulting from gum disease, even in those quite young.

As civilisation progressed, there developed a very marked difference between the diets of the wealthy and the poor in all countries and hence in their dental health. For example, in Egypt caries was rare among the common people until the time of the Ptolemies

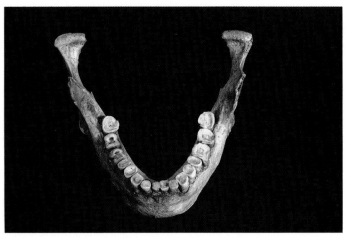

Fig. 1.1 The enamel on the biting surfaces has been completely worn away, exposing the pulps of the teeth (BDA Museum).

(323–30 BC), whereas from as early as 2500 BC the jaws of many aristocrats had caries and dental abscesses.

In Britain from the twelfth century onwards, sugar, recognised today as harmful to the teeth, gradually became more widely available, although it was still scarce and expensive. A German visitor to the court of Queen Elizabeth I described her as having black teeth which, he stated, were characteristic of the English from their excessive love of sugar. We can see from human remains over the centuries that, as the consumption of sugar rose, so did the level of caries in the population.

In the past, very many causes have been suggested for caries. For hundreds of years it was widely believed that caries and toothache were caused by a worm in the teeth. One of the earliest references to this belief occurs in a papyrus dating from at least 1000 BC. The theory was finally disproved in the mid-1700s, but it persisted to some extent into the present century.

Theories of the eighteenth and nineteenth centuries

During the eighteenth and early nineteenth centuries theorists suggested that decay started from the centre of the tooth (the pulp) as a result of inflammation and then proceeded outwards. Trouble with the teeth was frequently seen as a symptom of another disease.

According to others, caries started on the outside of the tooth and spread inwards, probably initiated by acids contained in the food. In fact, some authors realised that sweet, sticky food in the diet played a part in the causation of caries, although their explanations of how this happened were very far off the mark. Luxurious ways of life were often suggested as a cause of caries.

Another belief was that caries was caused by 'lateral pressure', that is, one tooth pressing against another. Those who supported this theory pointed out that if there were spaces between the teeth they seldom decayed, whereas teeth which were close together were much more susceptible to caries. What they did not realise was that these crevices retain plaque, which is the real villain.

Technical improvements in microscopes in the first half of the nineteenth century encouraged the study of the minute structure of the teeth in health and disease and produced the first new ideas for centuries on the complex causes of caries. In 1868 it was suggested that one particular bacterium, *Leptothrix buccalis*, was the cause of caries. This bacterium is now known to be a normal, and probably

Fig. 1.2 W. D. Miller (1853–1907), author of the influential *Micro-organisms of the human mouth*, published in Philadelphia, 1890.

harmless, inhabitant of the mouth, but the importance of the suggestion was the new idea that bacteria played a part in caries. Further research by W. D. Miller (fig. 1.2) led to the publication in 1890 of probably the most important work on the subject.[1] He showed conclusively that caries was due to the initial action of acids resulting from fermentation of food followed by the action of bacteria on the softened tissues. This is known as the chemico-parasitic theory and is still regarded as essentially accurate a century later.

Scientific research of the twentieth century

Twentieth-century research into the causes of dental disease has focused on trying to discover just which bacteria are involved and how they act. In the process, dental scientists have devoted much attention to what is now called 'plaque'. This is the pale yellow film which sticks to the teeth and is composed mostly of bacteria. One species of these (fig. 1.3) feeds on dissolved sugar, and the acid produced demineralises the enamel and dentine. Further bacteria then digest the organic remains. Bacteria also release toxins which cause inflammation of the gums, leading to the breakdown of the fibres and bone which support the teeth.

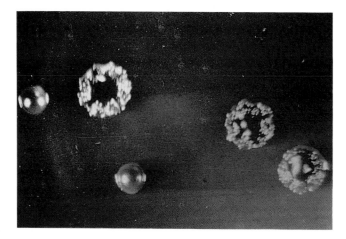

Fig. 1.3 *Streptococcus mutans*, the initial cause of tooth decay.

All this microscopic activity had been invisible to earlier generations of dentists. Writers of the eighteenth and nineteenth centuries mistakenly thought that 'tartar' was the actual cause of gum disease, bone loss and the loosening of teeth; they did not realise that, in fact, it was providing the perfect resting place for plaque, the active agent.

Diseases of the gums and tarter formation had been recognised for a long time. A large number of examples are recorded by old writers of persons who had one single 'bone' in a jaw instead of teeth. There is no doubt that the teeth were actually covered with a massive deposit of tartar. Most writers were cautious about providing an explanation of tartar formation, but they generally advocated its careful removal and directed that the teeth should be polished. They considered that gum disease was also caused by scurvy or the medicinal use of mercury.

2

Dental Treatment

With dental disease of one kind or another so rife throughout human history, one might expect that effective treatment would have developed at an early date, but this did not in fact happen. Before the early eighteenth century the type of treatment on offer was very limited in its scope and effect.

Cleaning the teeth

This has long been recognised in most societies not only as a desirable social habit but also as a means of keeping disease at bay.

In the East, the frayed ends of twigs of various bushes were, and still are, used. The instrument thus formed is known as a siwak (fig. 2.1). In Western society, teeth were usually cleaned with a piece of linen or sponge. Toothbrushes date from the mid-1660s. Among the Verney papers[2] is a letter written in 1649 to Sir Ralph Verney, then in exile in France, asking him to enquire in Paris for 'the little brushes for making cleane of the teeth . . . together with *petits bouettes* to put them in'. There are many eighteenth and nineteenth century examples of toothbrushes with gold or silver handles. Some of these

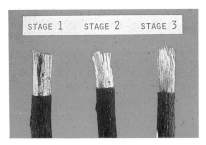

STAGE 1   STAGE 2   STAGE 3

Fig. 2.1 Stages of cutting and splaying a siwak ready for use.

Fig. 2.2 Set of toilet instruments, including toothbrush with replaceable head, toothpowder box and toothpick (BDA Museum).

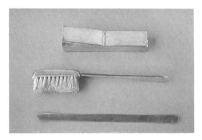

Fig. 2.3 Toothpowder box, toothbrush and tongue scraper in silver (BDA Museum).

Fig. 2.4 Toothpicks in a variety of materials (BDA Museum).

were part of elaborate toilet cases (fig. 2.2) or travelling sets which included toothpowder boxes and tongue scrapers (fig. 2.3).

Toothpicks date from very early times and were used in ancient China and other Far Eastern countries. The Roman poet Martial recommended that a toothpick should be made from mastic wood or, failing that, a quill. In the sixteenth and seventeenth centuries it was not uncommon for the wealthy to wear toothpicks of gold or silver, enriched with precious stones, on a chain around the neck; in the eighteenth and early nineteenth centuries they were carried in small decorated cases of gold, silver, ivory, tortoiseshell or wood. The toothpicks themselves were also of these materials (fig. 2.4).

Toothpowders also have a long history. Messalina, the notorious wife of Emperor Claudius, used a mixture of calcined stag-horn, mastic and sal ammoniac. Until quite recent times, toothpowders

Fig. 2.5 Lid from a pot of Gabriel's 'Royal Dentifrice' (BDA Museum).

frequently contained very abrasive substances such as finely powdered brickdust, china and earthenware, soot, cuttlefish bone or pumice-stone. Many dental practitioners marketed toothpowders made up to their own secret formulae (fig. 2.5). The few that were patented often contained very many ingredients; one product in the year 1800 was a combination of nine different substances.

Toothpaste was not widely used until the introduction, in the late nineteenth century, of soft collapsible tubes, made from a metal which did not affect the paste. Similar tubes had been used for artists' paints for a considerable time before this.

Some cures for toothache

Whether many people availed themselves of these preventive measures is another matter. The majority were probably more likely to be in search of a cure for toothache. Whenever and wherever they lived, they did not have far to look. Few ailments in the history of man can have prompted more 'remedies' than toothache.

In some societies, charms were worn to ward off its attack, in others fervent prayers and solemn chants were declaimed. In medieval Europe, urgent pleas were made to St Apollonia, the patron saint of toothache sufferers (fig. 2.6). There are many stories about why she holds this strange honour, but the following is fairly representative. In AD 249, in Alexandria, followers of Christ were set upon by violent mobs. Some gave in to the threats and worshipped the Roman gods, but a small number upheld their Christian beliefs. Amongst them was Apollonia, who was caught and imprisoned. In order to try to make her renounce her faith, her jaws were crushed and her teeth extracted, one by one. As she stood in front of the funeral pyre, Apollonia was

Fig. 2.6 St Apollonia sits reading, holding a pair of forceps with an extracted tooth in her right hand (MS Astor A.24, vol. 2, fol. 82v. Reproduced by permission of Lord Astor).

given the choice of returning to the old gods or perishing in the fire. In reply, she jumped into the flames to die the death of a martyr. In spite of the pain, Apollonia prayed that in future, anyone who invoked her name would receive relief from toothache. One of the prayers used by sufferers ran:[3]

> O holy Apollonia, intercede for us by thy passion,
> by thy suffering in the teeth, throat and tongue,
> that we may be delivered from pain in the teeth now
> and for ever.

A more pagan form of magic lies in the advice, 'Take a new Nail and make the Gum bleed with it, and then drive it into an Oak'. Perhaps the same principle of exorcising evil was behind the advice to apply the body of a mouse to the aching tooth.

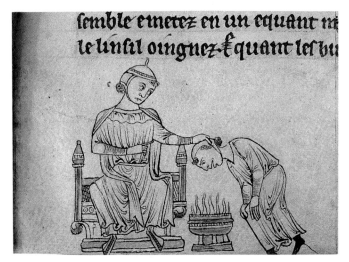

Fig. 2.7 Fumigating the mouth to drive out the 'toothworms', in reality henbane seeds which the operator had concealed in his hand. (MS 0.1.20. Reproduced by permission of the Master and Fellows of Trinity College, Cambridge).

Some remedies contained a grain of sense: applying warm poultices and mouthwashes would certainly have eased the pain. Signieur Francisco Dickenson recommended, in 1649, that the sufferer 'take a little Cotton and dip it into essence of Cloves' and place it 'in the hollowness of your tooth';[4] oil of cloves is still used in a temporary filling material today. An ancient Egyptian contribution to the dental pharmacopoeia reads: 'Cow's milk 1 part, fresh dates 1 part, and uah corn 1 part. This is to be left to stand and then masticated nine times.'

The Romans advocated narcotics and astringents. In societies where caries was put down to toothworms, the obvious treatment was to get rid of those worms. This feat was performed by sleight of hand by quacks in the market place (fig. 2.7): the operator concealed on his person maggots or pods of henbane seeds which, when heated, burst open and produced seeds then claimed as toothworms which had come out of the mouth. More effective remedies, though demanding heroism on the part of the patient, were cauterising the nerve of the tooth with a red-hot iron or lancing an abscess.

Extraction

Surprisingly rare at this time was the extraction of the offending tooth. In the ancient world, this was an operation undertaken with reluctance, partly because of the pitfalls awaiting the unwary operator of inadequate instruments. By the Middle Ages, however, it appears to have become somewhat more common. When the remains from the *Mary Rose* were examined, for example, it was clear that the unfortunate sailors had had teeth extracted.

In the seventeenth century, extractions were often performed with the patient seated unceremoniously on the floor (fig. 2.8).

The more technical approach

Attempts to actually repair the ravages of caries and gum disease were rare. Despite the sophistication of their society in other directions, the remains of ancient Egyptians show no signs of fillings and there is no firm evidence of attempts to replace missing teeth with false ones, despite claims to the contrary. A very few examples have been discovered of loose teeth bound together by gold wire, but this seems more likely to have been carried out during burial preparations. The

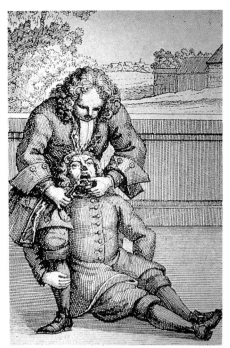

Fig. 2.8 Tooth extraction in 1717, depicted by Cron (Guerini, V. *A history of dentistry.* New York, 1909).

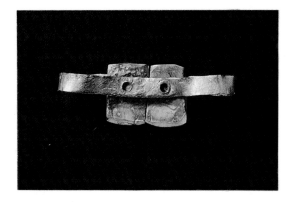

Fig. 2.9 An Etruscan denture (Reproduced by permission of Liverpool Museum).

aim was to keep the body intact for the next world, rather than to restore the usefulness of the teeth for this one.

One of the peoples of antiquity who did attempt dental prostheses (false teeth) were the Etruscans who lived during the seventh century BC in what is now Tuscany. They were already very skilled at working bronze and used this experience to make narrow bands (3–5 mm wide) of highly flexible, very pure gold to put around loose teeth and so splint them together. Sometimes they formed these bands into a framework to which an artificial tooth made from ox tooth was riveted (fig. 2.9). These are the first real attempts at reconstruction in dentistry. It may have been a status symbol to display gold in the mouth in Etruscan society; cultures vary quite markedly in their ideas of aesthetic excellence.

In smart Roman society, if the poet Martial is to be believed, prostheses like the Etruscan ones were not unusual but few of these gold pieces have survived. This is partly because of the widespread practice of cremation and partly because, in a period of raging inflation, cheaper substitutes were apparently sought, providing the satirist with further targets:[5]

> Aegle feels she still has her teeth because of her
> purchased bone and eastern ivory
> At night she takes out her teeth just as she takes
> off her finery.

The removable denture had made its début and so had ivory, not to be superseded as a material for denture bases for nearly 2000 years.

During the Middle Ages these techniques seem to have been largely lost. The surgeons of the sixteenth and seventeenth centuries, with their new interest in anatomy, became more adventurous in their treatment of such conditions as broken jaws, perforated palates and tumours in the mouth, but they did not concern themselves greatly with the individual teeth. Nevertheless, in the seventeenth century, it became more common for repairs to be made to decayed teeth and for missing ones to be replaced. The initiative came less from the medical world than from other quarters.

The dazzling French court of Louis XIV (who reigned from 1643 to 1715) had attracted people from all over Europe as it became a centre for extravagant and luxurious living. Goldsmithing and the decorative arts reached new heights. In such a society, the fashionable élite were anxious to keep their youth and looks, which was difficult if every elegant smile revealed great gaps or a mouthful of decayed teeth. Those craftsmen who catered for their needs and whims responded to the call and, as a result, Paris became renowned for the excellence of its dental treatment.

Dental treatment during the eighteenth and nineteenth centuries

It was during this period of remarkable social and technological innovation that the foundations of modern dentistry were laid. In the eighteenth century the first real textbooks on the subject appeared. This marked a startling departure from previous practice, whereby techniques had been jealously guarded and only the foolhardy dentist divulged his secret methods. Now, as dentists vied with each other to prove they were every bit as good as, if not better than, the next one, new remedies for old problems emerged and the efficiency of long-tried methods was improved. Not that all was perfect: quackery was as rife as ever in a society where knowledge was expanding rapidly while most people remained uneducated.

Treating toothache and loose teeth

Jiggling the tooth to 'slacken' the nerve was an attempt to cure toothache on the grounds that the nerve could not then transmit any sensations of pain. For home treatment, little 'anodyne' pills were

Fig. 2.10 A pelican (BDA Museum).

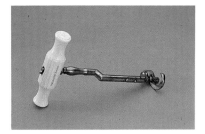

Fig. 2.11 A key (BDA Museum).

sold to place in a cavity. As these contained opium, they would at least deaden the pain for a while.

Gum disease was treated by very thorough cleaning of the teeth and gums with extractions only as a last resort. This sometimes involved the use of 'tinctures' (in reality strong acids) which, as well as dissolving the tartar, also made short work of the enamel of the teeth. Loose teeth could be tied to nearby firm ones with finely annealed gold wire, of benefit in the short run, perhaps, but likely to lead to further loss of teeth in the long run. Reimplantations were attempted after the teeth had been knocked out in an accident, especially in the case of children.

Extraction instruments

In the eighteenth century, extractions were carried out with a range of barbaric-looking instruments, including the gum lancet, punch, forceps with curved beaks and handles, levers and pelicans. The last were so called from the resemblance of the claw to the beak of the bird (fig. 2.10). Growing in popularity was an instrument called, again for reasons of its shape, the 'key'. The action of both the pelican and the key depended on leverage against the gum and there was a very great likelihood of extensive damage to the gums and jawbone. There was also a danger of neighbouring teeth being accidentally extracted at the same time (figs 2.11 and 2.12).

The nineteenth-century contribution to the art was the development of forceps which had beaks anatomically shaped to fit the crowns of the teeth (fig. 2.13). These were introduced in an effort to reduce slippage of the instruments on the teeth and to stop the terrible damage to the surrounding gums and bones that had been so common with the pelican and the key.

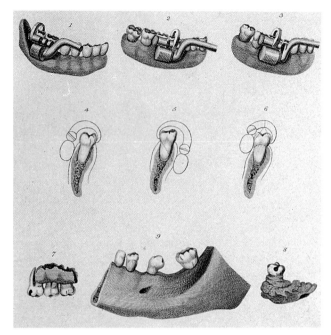

Fig. 2.12   Use of the dental key (Fox, J. *The history and treatment of the diseases of the teeth*. London, 1806).

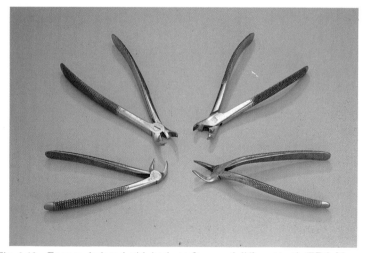

Fig. 2.13   Forceps designed with beaks to fit around different teeth (BDA Museum).

Making good the damage

Artificial teeth

In the eighteenth century the materials for both teeth and denture bases varied, but were predominantly of natural origin. The most favoured for a denture base (the part to which the teeth are attached) was hippopotamus ivory (fig. 2.14). Into this would be riveted human teeth. Potential clients were often reassured that these came only from reliable sources, that they were 'Waterloo teeth', taken from healthy young men slain on the battlefields. In reality, of course, most came from the mortuaries, the dissecting rooms and from bodies snatched from the grave by resurrectionists.

To produce an accurately fitting denture, a model of the patient's mouth is needed so that the appliance can be fitted to it. In Prussia, in the middle of the eighteenth century, the technique was developed of taking an impression of the patient's teeth in wax and then using this to cast a model in plaster of Paris. This clever use of materials was surprisingly slow to spread to the rest of Europe. The dentist usually had to measure the mouth with compasses or outline the shape of the jaw with a piece of card. Then he completed the carving of the surface of the denture which fitted over the gums at the chairside.

Because it was so difficult to carve ivory to fit accurately over the whole of the palate, most complete upper dentures were horseshoe shaped, and such a denture will rarely stay in place (fig. 2.15). To try to solve this problem, springs were designed which joined the upper denture to the lower one; their action was to force the two pieces apart

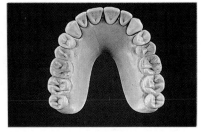

Fig. 2.14 A denture made from hippopotamus ivory is carved to fit a plaster model of the patient's mouth (BDA Museum).

Fig. 2.15 A typical horseshoe-shaped, ivory upper denture (BDA Museum).

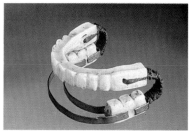

Fig. 2.16 Ivory full denture with spiral springs to keep them in place (BDA Museum).

Fig. 2.17 Facsimile of early eighteenth century ivory upper denture retained in the mouth by a framework fitting over the lower teeth (BDA Museum).

and onto the gums (fig. 2.16). Early in the eighteenth century, the springs were made of whalebone strips, but these were gradually replaced by steel ones whipped with thread, and finally by spiral springs made from gold or silver. Where there was no lower denture to attach the springs to, ingenious gold frameworks were devised to fit around the standing lower teeth and take the other end of the springs (fig. 2.17).

Partial dentures (to replace only a few teeth) were generally quite small and neat. They were not designed to be removable and so were very difficult to clean. Indeed, a very real problem with all these early dentures was the lack of hygiene. It was a source of great dismay that both ivory bases and human replacement teeth decayed as rapidly in the mouth as did natural teeth. As a contemporary dentist admitted:

To say nothing of this contaminating putrid accumulation being carried to the stomach of the unfortunate wearer, the unavoidable thickness, the liability to discoloration, decay and the expense of renewing, are insurmountable objections to the use of human teeth set on a base of bone.

No wonder the use of the fan became a social necessity! (fig. 2.18.)

Porcelain, a new hope

Towards the end of the eighteenth century, the general dissatisfaction with ivory stimulated experiments with porcelain and the production of one-piece 'incorruptible' dentures (fig. 2.19). True, these were a

Fig. 2.18 Fans in use in the eighteenth century (Reproduced by permission of the Victoria and Albert Museum).

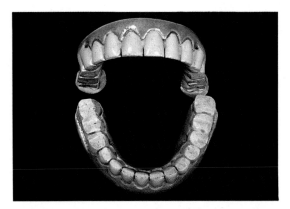

Fig. 2.19 An early specimen of complete upper and lower porcelain dentures of French make. Provision has been made for the use of springs. (Reproduced by permission of the Odontological Museum, Royal College of Surgeons of England).

Fig. 2.20 De Chémant as depicted by Rowlandson, 1811 (11798 in the *British Museum Catalogue of Political and Personal Satires*).

great improvement in terms of hygiene but they were all too brittle and very often the colours and contours were crude. They were also the subject of a good deal of hilarity at the time (fig. 2.20). Their eventual demise, however, was due to the difficulty of controlling shrinkage during firing and of producing a reasonably good fit.

The work on porcelain as a material for dentures was not entirely wasted, however. In the early nineteenth century, single porcelain teeth began to be made with attached platina brackets which could be soldered to a gold or platina base. These 'terro-metallic' teeth were initially bean-shaped and crude but because they did not decay they were certainly a distinct advance on human replacements (fig. 2.21). Far more refined were the 'tube teeth' produced by Claudius Ash of London in the late 1830s. These had a central sleeve of gold or platinum which slipped over a post attached to the denture base; they could even serve as a simple form of crown when attached to a post fixed into a root (fig. 2.22). All-in-one blocks of teeth made of porcelain were later developed for use at the back of the mouth where function was more important than appearance (fig. 2.23). A pink glaze could be fired on to a platina base around the necks of the teeth to provide a natural appearance, but such dentures proved very fragile in use.

Fig. 2.21 Terro-metallic teeth (BDA Museum).

Fig. 2.22 Tube teeth produced by Claudius Ash (BDA Museum).

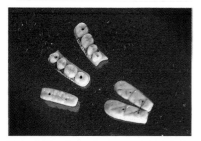

Fig. 2.23 One-piece porcelain blocks for fixing in the molar area of a denture (BDA Museum).

Fig. 2.24 Cast zinc swaging block for gold denture plate (BDA Museum).

Fig. 2.25 A partial denture with gold base and porcelain teeth (BDA Museum).

Once models of teeth and gums came into wider use, it became possible to eliminate ivory altogether by making the denture base out of gold. To do this, a gold sheet was swaged down on to a duplicate model made from zinc (fig. 2.24). To keep partial dentures in place, wire clasps could be fitted around the remaining natural teeth (fig. 2.25). These early metal bases fell into disrepute: patients felt they were not 'natural', a collapsed upper lip could not be plumped out as with ivory and they were uncomfortable on the crest of the gums. Not infrequently, they corroded badly because unscrupulous dentists used pinchbeck or gilded brass instead of gold.

The search for alternative materials for bases went on; tortoiseshell and gutta-percha were tried, but with little success. So, it was back to ivory, with all its associated problems.

At last, the denture breakthrough

The introduction of anaesthesia to dentistry (see p. 27) had a dramatic effect on the numbers of those seeking treatment. People were prepared now to have teeth extracted rather than grimly hanging on to rotting roots. The demand for false teeth increased the pressure to find some viable alternative to ivory that was less expensive than gold. This was at last realised in the 1850s with the development of denture bases made out of vulcanite and fitted with porcelain teeth. Unfortunately, the cheapness of vulcanite and the ease with which it could be made upset those who saw their skill and craftsmanship being usurped by 'mere technicians' (fig. 2.26).

Denture construction was made even simpler when it was realised that plaster could also be used to take impressions, giving much more detail of the mouth than wax. Gutta-percha and a material known as

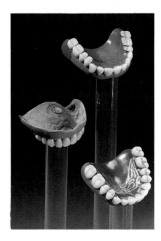

Fig. 2.26 Vulcanite dentures, one fitted with a suction device to aid retention (BDA Museum).

'Stent's compound' were introduced as impression materials at this period. It had long been realised that if the denture completely covered the palate it stayed in place far better because of the greater suction afforded. These better impression materials made this a reality.

New crowns for old

Dentures were not the only way of replacing missing teeth. Transplantation was widely practised for a while, particularly at the end of the eighteenth century. Poor young girls and boys were paid to have their healthy front teeth removed for immediate placement into the waiting jaws of wealthy, older patients who had just had their blackened stumps extracted (fig. 2.27). Long-term success was extremely rare.

Where the crown of a tooth had decayed, leaving the root behind, it could be replaced by the crown of a human or animal tooth, attached by a gold pin forced into the root canal (fig. 2.28). Gold shell-crowns were made in the mid-eighteenth century to cover severely worn-down teeth. If a few teeth were missing, the replacements could be

Fig. 2.27   Transplanting of teeth, depicted by Rowlandson, 1787 (7766 in the *British Museum Catalogue of Political and Personal Satires*).

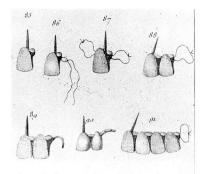

Fig. 2.28   Crowns and bridges retained by pins and ligatures (Laforgue, L. *L'art du dentiste*. Paris, 1802).

Fig. 2.29   Partial denture of ivory to be attached to adjacent teeth by silk threads (BDA Museum).

fastened to the remaining natural teeth with silk or gold wire (fig. 2.29), or even suspended like ear-rings from pins which had been inserted into the gums. Towards the end of the nineteenth century, all-porcelain crowns were developed for cementing on to a natural root.

Artificial palates (obturators)

Syphilis was one of the great scourges of pre-antibiotic society. Especially if treated with mercury, it can lead to the destruction of the bone and soft tissues of the palate and nose. This renders sufferers unable to speak or eat. There were also those unfortunate people born with cleft palates.

The simplest remedy was to try to plug the gap with a piece of linen or a sponge but these soon became unpleasant and could actually make the hole larger as they swelled on contact with moisture in the mouth. Some dentists attempted to get the hole to close up; they scraped the edges to encourage them to knit together as they healed. Often the tissue destruction or defect was too great for this and a plate had to be made.

Much ingenuity went into creating these artificial palates to keep nose and mouth separate. A butterfly design was produced in ivory; the 'wings' were passed through the cleft into the cavity of the nose and then opened out with the help of a key (fig. 2.30). Others were kept in place by threads which, passing through holes drilled in the

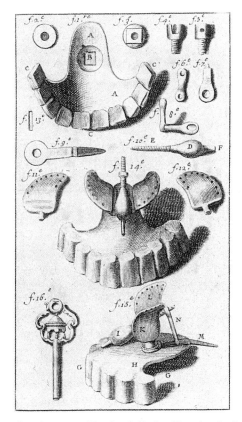

Fig. 2.30 Designs for obturators (Fauchard, P. *Le chirurgien dentiste*. Paris, 1746).

appliance, were then tied to any remaining teeth. Drainage holes could be provided for saliva or mucus to escape. Pieces of sponge were sometimes attached but, of course, these absorbed mucus; the result was that hygiene was little better than in the days of linen plugs.

The introduction of vulcanite in the 1850s was a boon to those in need of an obturator, for this material could be fitted to the mouth so much more accurately than previous materials.

Filling the teeth
In the eighteenth and nineteenth centuries caries in its early stages was frequently filed away, with the intention of leaving a smooth

surface. Deeper caries was removed by cautery or was scraped out. The cavity was then filled with tin or lead. Earlier generations of dentists had used wax or mastic and coral on the rare occasions when they filled a tooth. The most suitable material available was gold foil, rammed hard down into the cavity, but this was, of course, expensive.

In the early nineteenth century a number of cheaper substitutes were tried, including mercury amalgams based on heavy metals or silver filings from, originally, French coins. Pellets of amalgam were put into the cavity and worked with a hot instrument. Some had such a low melting point that they were simply poured into the cavity in molten form and left to solidify. Poor mastery of the technique and unscrupulous operators gave these amalgams a bad name. In fact, the American Dental Association went so far as to try to get all its members to sign a declaration that they would never use them. The furious disagreement over their value and safety came to be called the 'Amalgam War'; not until the end of the century were practitioners convinced that amalgam was suitable as a filling material.

During the nineteenth century, the technique of filling root canals was further developed. (This can sometimes be done in a case where decay has exposed the central pulp. The canal is cleaned out and filled and the tooth restored by a filling or crown.) At first the materials used were cohesive gold, cotton fibres, wood or asbestos. In the middle of the century gutta-percha (still appropriate today) and gold wire were used; by 1900, root treatment involved creosote, carbolic, iodoform and various formalin cements. The crowns that were frequently fitted over these root fillings had improved in design so that they stayed on better, were more natural in appearance and did less damage to the surrounding gums.

Before the 1870s, when the dental engine was introduced, all this was done with hand instruments; the 15-minute dental appointment was unknown. Early patients are to be admired for their powers of endurance—it is not surprising that there were so few of them. There was no concept of sterilisation at this time, before the discovery of the role of bacteria; the handles of instruments were often ornately carved (fig. 2.31) making them liable to harbour infection.

Orthodontics

One aspect of modern dentistry which owes much of its early development to this period is orthodontics—the moving of mal-

Fig. 2.31 Dental instruments with elaborate mother-of-pearl handles, c. 1860 (BDA Museum).

aligned teeth for health or cosmetic reasons. There was a very ancient tradition of applying daily finger pressure to guide a misplaced tooth into a better position, but little of note was written about correction of irregular teeth until the eighteenth century. Joseph Fox was, in 1803, the first English dentist to publish explicit instructions for the correction of these irregularities;[6] he was the first to classify them and his methods were followed for many years (fig. 2.32). Several French practitioners invented new appliances in the first decades of the nineteenth century, including some (the inclined plane and the labial arch) still in use today. The rubber band, familiar to the present-day orthodontic patient, first made its appearance at this period.

Interest in orthodontics spread rapidly in the United States, although it was not always systematically carried out. From the confusion emerged one of the most influential figures in dentistry, E. H. Angle, who qualified in 1876 (fig. 2.33). He had become critical of the mere gadget-minded methods still characteristic of many practitioners in a rapidly expanding field and tried, unsuccessfully, to have increased time for orthodontic education in dental schools. He published his classification of malocclusions (deviations from the normal position of the teeth) and reported his work at a meeting in 1899.[7] His audience received him so enthusiastically that he gave a course on his methods in his practice. This side of his work snow-balled and resulted in the Angle School of Orthodontia in St Louis. The scientific quality of its dogmatic teaching and the calibre of its

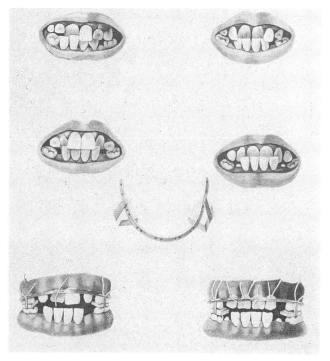

Fig. 2.32   Designs for orthodontic appliances (Fox, J. *The history and treatment of the diseases of the teeth*. London, 1806).

Fig. 2.33   E. H. Angle (1855–1930), an influential figure in dentistry (BDA library collection).

graduates had an enormous effect on the course of orthodontics for many years.

General anaesthesia
The nineteenth century saw substantial improvements in dental treatment, but few can have been so beneficial to mankind as the introduction, by dentists, of anaesthetic gases.

The prospect of pain loomed large in the imagination of the reluctant dental patient during the first half of the nineteenth century. This was also a harrowing problem for surgeons but their patients, usually desperately ill, had little alternative to accepting the agony inflicted. Both surgeons and dentists relied on operating as rapidly as possible, with their patients sometimes under the influence of a judicious prescription of alcohol or opium.

Anaesthesia has its roots in the discovery of nitrous oxide by Joseph Priestley in 1772. This gas was found by Humphrey Davy and others to have an analgesic and intoxicating effect (it was popularly called

Fig. 2.34 Laughing gas party, illustrated by Cruikshank (Scoffern, J. *Chemistry no mystery*. London, 1839).

Fig. 2.35   Horace Wells (1815–48). Portrait painted by Charles Noel Flagg (Reproduced by permission of the Wellcome Institute, London).

'laughing gas') (fig. 2.34). A young American dentist, Horace Wells (fig. 2.35), became convinced that it could also render patients unconscious and used it effectively on several occasions. Unfortunately, a demonstration he gave at the Massachusetts General Hospital in January 1845 proved unsuccessful and, not surprisingly, met with a poor reception. Wells became depressed and never recovered from his apparent failure.

His associate, William Thomas Morton, was at the same time testing ether which also induced unconsciousness. After experiments on animals and on himself, Morton successfully anaesthetised a patient and extracted an infected tooth on September 30, 1846. On October 16 he demonstrated his method at the Massachusetts Hospital when a patient had a tumour on the neck removed. This historic occasion has been accepted as the birth of surgical anaesthesia (fig. 2.36).

The first British person to give an anaesthetic (on December 19, 1846), only two days after the momentous news reached this country by letter) was also a dentist, James Robinson, of Gower Street, London (fig. 2.37). A commemorative plaque marks the location of this event.

As the use of anaesthetics was explored further, it soon became clear that neither ether nor chloroform (introduced in 1847) was ideal

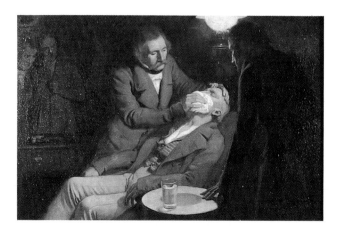

Fig. 2.36   The first use of ether in dental surgery, 1846. From a painting by E. Board (Reproduced by permission of the Wellcome Institute Library, London).

Fig. 2.37   James Robinson (1816–62). (From Cock, F. W. The first major operation under ether in England. *American Journal of Surgery* (Anesthesia supplement) 1915; **29**: 98–106).

for dental operations and several dentists turned again to nitrous oxide. By the 1860s it was being used in America and in France from where it was reintroduced into Britain by Thomas W. Evans. Joseph Clover later introduced a nasal mask which permitted anaesthesia to be given throughout the operation instead of only before it. Mixing the gas with air and at a later date with oxygen enabled anaesthesia to be prolonged.

By the end of the nineteenth century dentistry was becoming increasingly technical. Gone were the days when all a dentist needed was a small case of instruments, a table and a chair. The recognisable dental surgery had emerged, with its specialised operating chair, dental engine and cabinets (fig. 2.38), all produced by dental manufacturers and suppliers. As knowledge began to be widely available in professional journals, on both sides of the Atlantic, so clinical and laboratory techniques were put on to a more scientific basis. The principles laid down in the 1890s for preparing a tooth to receive a filling are only just being superseded. Dentures were no longer made just to fit together, but placed on devices called articulators which attempted to mimic the movements of the jaws, ensuring that the finished product would perform more realistically.

Fig. 2.38   Reconstruction of a dental surgery of the 1890s (Science Museum).

Dentistry was no longer brutal butchery, nor was it merely an art. It was now well placed to take advantage of the explosion in new materials and techniques which was to follow.

Twentieth-century treatment

If the introduction of general anaesthesia was the godsend of the nineteenth century, that of the twentieth must surely be the better scientific understanding of dental disease and how to prevent it. Not that this picture is complete; for instance, although we now know the names of some of the bacteria thought to be responsible for gum disease, we are little further forward in discovering why some people are more susceptible to the condition than others. A fully effective vaccine against harmful bacteria in the mouth has yet to appear.

However, twentieth-century dental research has produced very positive observations which are directly applicable to clinical treatment and the care of the teeth. For example, it is now known that caries is related not to how much sugar we eat, but how often we eat it. This makes it possible to advise patients on how to change their eating habits to cut down the risk of decay.

Another modern preventive measure is based on the observation made early this century that people who live in areas where there is fluoride occurring naturally in the water supplies have less tooth decay. Scientists have since discovered that the fluoride combines with the hydroxyapatite which makes up 97% of the tooth enamel. The result is fluorapatite, a mineral which is less soluble in the cocktail of organic acids of plaque than plain hydroxyapatite. The obvious follow-on to these findings, the addition of fluoride to water supplies on a widespread scale, has been delayed (for political rather than scientific reasons) but over 95% of toothpaste sold in this country now contains fluoride. This has helped to reduce caries in children since 1965 by about 40%. Dentists, dental hygienists and dental therapists also apply fluoride solutions and gels to the surfaces of the teeth to help prevent decay.

As far as gum disease is concerned, much more is now known about the changes that occur within the tissues when the gum becomes inflamed. This has led to greater success with the treatment. On the preventive side, in 1965 it was proved that plaque on the teeth is the most important cause of gum disease. Its daily removal by the patient is more important for the health of the gums than the scaling of the teeth by the dentist.

Despite a fall in the caries rate in the young, there is still a need for restorative work, particularly in the generation who have not had the maximum benefits of fluoride toothpaste and already have fillings. Restorations have a limited life and each time they are replaced, the task is likely to be more complicated. This has stimulated considerable advances in techniques and dental materials. Until recently, tooth-coloured filling materials (based on silicates) had been suitable only for front teeth, but 1985 saw the introduction of the latest varieties for use at the back of the mouth. Most white filling materials are now composed of an acrylic resin to which has been added a carefully graded, fine silicaceous powder. These 'composite' fillings can now be bonded directly to the tooth surface. When amalgam is used, an extensive area of the tooth often has to be cut away to provide a box with suitable undercuts to prevent the filling coming out again.

In the early years of this century, the ancient technique of casting was applied to dentistry. Gold crowns, bridge work and, particularly, inlays could now be produced (fig. 2.39). Gone were the hours of enduring gold foil being rammed into a cavity with a mechanical mallet. Gold is now less used as a filling material, on grounds both of cost and aesthetics. It is confined to special uses, as, for example, in crowns to which porcelain is bonded.

Until the 1940s, vulcanite remained the chief material used for denture bases. Although very serviceable (there are still a few patients around wearing vulcanite dentures made anything up to 60 years ago), it did not score highly as regards colour (reddish-brown) or hygiene. Alternatives were already being sought during the 1930s and the shortage of rubber from the Far East during World War II pushed acrylic resin to the fore as an alternative denture material. It is

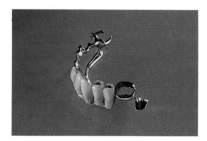

Fig. 2.39   Cast gold 'Stressbreaker' denture, 1930s (BDA Museum).

Fig. 2.40 Acrylic dentures (BDA Museum).

still the most widely used, both for the base and the teeth (fig. 2.40).

Much of the twentieth-century treatment has been made possible or easier by the improvement in equipment. The turbine handpiece, with its air bearings (developed in the late 1950s), probably had an even more dramatic effect than the introduction of the electric motor. The bur is driven at up to 400 000 revolutions per minute and so, for the first time, enamel and dentine could be cut very quickly, accurately and with the minimum of vibration and discomfort. This was an inconceivable experience for the patient of a century ago.

Happily, this coincided with the more widespread use of local anaesthesia. This had its origins in the 1880s, when Carl Koller of Vienna discovered that cocaine could be used as a surface anaesthetic to block out the sensation of pain. Many American and some British dentists, however, began to administer cocaine by injection. This was an important milestone for dentistry, since the general anaesthetic gases of the nineteenth century could be used only for extractions and similarly short operations. Fillings and other more lengthy treatments had remained an agonisingly painful experience.

Unfortunately, it soon became clear that cocaine could produce toxic side-effects and damage the tissues of the mouth. Chemists in Germany set about finding a way round this and succeeded in synthesising a new drug called 'procaine' in 1905. Although this could produce allergic reactions, it remained the main local anaes-

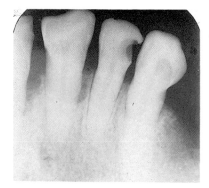

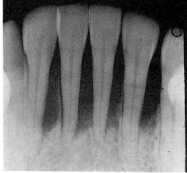

Fig. 2.41 Radiograph showing caries and bone loss.

Fig. 2.42 Bone loss so severe that the teeth are barely supported in the jaw.

thetic for a further 40 years. Since about 1950, the more acceptable lignocaine has been the drug of choice. The importance of local anaesthesia cannot be overestimated; it opened up whole new possibilities in dental treatment. Without it, dentistry as we know it today would be unthinkable. More recently, general anaesthetics and sedatives, given directly into a vein, have been developed for restorative treatment in certain circumstances.

A new era was opened up by the discovery of x-rays by Röntgen in 1895; one of its first clinical applications was in dentistry. For the first time, dentists could visualise what lay beneath the surface without recourse to surgery. In addition to showing the presence and position of unerupted teeth, radiographs can help the dentist to reach a diagnosis by showing such features as the extent of caries in an individual tooth, the presence of infection progressing into bone from the teeth or the amount of bone loss by periodontal disease (figs 2.41 and 2.42).

Finally, some treatments now available are unique to the twentieth century: computer imaging to forecast the outcome of orthodontic cases, the reshaping of teeth with composite resins, and sealing the fissures of children's teeth with acrylic resin to prevent their decay. A far cry indeed from the ivory denture and lead fillings of 200 years ago!

3

The Providers of Dental Treatment

As we have seen, any treatment carried out in the mouth and on the teeth was of a very limited nature until relatively recently. Toothache remedies were probably compounded by the herbalist or the apothecary and the very few prosthetic appliances fell within the province of the goldsmith. However, as caries became more widespread from the middle ages onwards, attitudes towards extraction appear to have changed and it became more frequently performed. In England, the barber often wielded the forceps but there also emerged a group of specialist toothdrawers, sometimes travelling the country wearing distinctive hats and strings of teeth (fig. 3.1).

Such men occasionally find their way into contemporary literature or official records. Some were granted licences to practise by the Company of Barber Surgeons of London or by their local bishop. For others, their place in posterity was achieved under less happy circumstances: in 1426 one Richard Feryer of Colchester was sued for damages of £40 by Richard Jeull and his wife Joan. It was alleged that Feryer had made an assault on Joan with his forceps, lacerating her tongue and 'wickedly drawing blood'. Feryer defended himself, claiming that he had treated Joan 'properly and according to his art'.

Feryer's surname is of some significance here since it was not unknown for blacksmiths, even into the nineteenth century, to turn their hand to manipulating the extracting instruments they made for the toothdrawers (fig. 3.2).

The 'operators for the teeth'
The seventeenth century saw the beginnings of change in the scope of dental treatment and the emergence of a new kind of practitioner, the 'operator for the teeth', defined in a contemporary dictionary as 'one

Fig. 3.1 A sixteenth century tooth-drawer, drawn by Lucas van Leyden, 1523 (R. A. Cohen's collection).

Fig. 3.2 The eighteenth century blacksmith/toothdrawer, a common subject for caricaturists. Drawn by Dixon, after Harris, 1768 (R. A. Cohen's collection).

skilled in drawing teeth and in making artificial ones'. It is not known how many of these operators there were nor whether they existed in any numbers outside London. Certainly they found a market for their services in the sophisticated capital where some, Thomas Middleton and Peter de la Roche, received royal appointments. Despite this regal recognition, the provision of artificial teeth was still far from widespread. The unfortunate Mrs Samuel Pepys, a martyr to tooth decay, repeatedly had teeth extracted by eminent toothdrawers and barber surgeons during her brief life, but her husband's diary makes no mention of any artificial replacements.

The early operators appear to have sent out metalwork required in their prostheses to be made by a goldsmith. By the end of the seventeenth century the London press carried advertisements placed by such men:[8]

Mr Pilleau a French Goldsmith living in St Martin's lane . . . does give Notice, that by an experience of 18 Years, he has found out a way to make and set Artificial Teeth in so firm a manner, that one may chew with them, ... Any Operator for the Teeth may buy some ready made from him.

A year later, Pilleau's advertisements omit any reference to supplying others and it has been surmised that by then he had become an operator himself, dealing directly with clients. Certainly there was nothing to prevent him from doing so.

In the early eighteenth century, the number of advertisements in the London press for operators increased. Clearly, the demand for this new fashion was growing.

The arrival of the dentist

A new word entered the English language in the late 1750s—'dentist', borrowed from French and the most usual form of description from then on for one concerned with matters dental. The range of dentists' work extended to restorative techniques and the treatment of gum diseases; they attempted to cover, in fact, very much the same ground as their modern counterparts, but without the benefit of a scientific understanding of the problems confronting them, of anaesthetics, or of modern technology.

Although demand seems to have risen for the dentist's services, no doubt with the increasing consumption of sugar, the market was still a limited one. Dental treatment was expensive (full dentures often cost 20 guineas in the mid-eighteenth century, the equivalent of about £500 in today's money) and not even all the rich were interested in caring for their teeth. It was a common practice from the middle of the century for London dentists to take to the road for a few weeks, advertising their services in the provincial press (fig. 3.3).

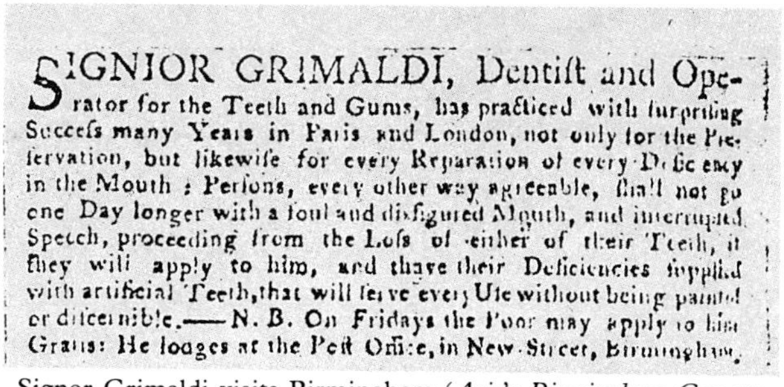

SIGNIOR GRIMALDI, Dentist and Operator for the Teeth and Gums, has practiced with surprising Success many Years in Paris and London, not only for the Preservation, but likewise for every Reparation of every Deficency in the Mouth : Persons, every other way agreeable, shall not go one Day longer with a foul and disfigured Mouth, and interrupted Speech, proceeding from the Loss of either of their Teeth, if they will apply to him, and share their Deficiencies supplied with artificial Teeth, that will serve every Use without being painful or discernible.——N. B. On Fridays the Poor may apply to him Gratis: He lodges at the Post Office, in New Street, Birmingham.

Fig. 3.3 Signor Grimaldi visits Birmingham (*Aris's Birmingham Gazette*, June 18, 1759).

The provinces began to produce their own dentists (the earliest yet discovered is Birch Hesketh in Liverpool in the late 1760s) who in turn found themselves taking to the road in search of patients, particularly at times when large groups of people were brought together for race meetings, an Assembly, the Assizes or the visit of a travelling theatre company. The amount of time spent on the road by some of these early dentists could be considerable. A Leicester dentist, James Blair, regularly embarked on at least two such tours a year, leaving his wife in charge of his perfumery and fancy goods shop. Each trip lasted several weeks and took in a number of places. In 1790, for example, he made a tour of the Manchester area in April, was in the Chester district for June, during which time he paid a subsidiary visit to Wrexham, and concluded the year with Cambridge in November. He maintained this pattern for the best part of 25 years, taking rooms at inns where he would see patients or visiting clients in their own homes. While the dentist's equipment could still be carried in a small bag, this was perfectly feasible.

Blair was one of a small band. Even in 1800, there were not many more than twenty 'dentists' in the provinces, with perhaps twice that number in London. The toothdrawers were also still in operation (fig. 3.4), although the growing band of chemists and druggists were taking over their function. Country doctors performed the occasional extraction, but seem rarely to have ventured further into dentistry.

These early dentists, whether in London or the provinces, came from a wide variety of backgrounds. Some emerged from the ranks of the watchmaker or goldsmith, others from the world of the hair-dresser or patent medicine vendor. There was no one to give these enterprising men a proper training; they were self-taught or had picked up hints from the advertising leaflets of their rivals. Subsequent generations considered them very ignorant and so, of course, they were by later standards. However, some managed to stay in practice for considerable lengths of time; in the harsh world of eighteenth century competitive trade, it is hard to believe that they would have managed to stay in business if their treatment of every case proved a disaster.

Taking on a newcomer for a few weeks' initiation soon became a profitable sideline for some dentists. A short 'course' such as this continued to be a common route into dentistry for a long time to come. However, even during the eighteenth century, a small number

Fig. 3.4 The mountebank or marketplace operator, 1771 (printed by Carrington Bowles, London).

of men were taking on apprentices and giving them a full four or five years' training, although there was no professional body to oversee the content of the instruction or provide a qualifying examination.

Apprenticeship became more common with time. A. J. Woodhouse has left us this description of his years as an apprentice in Exeter in the 1840s:[9]

The mode of instructing me that Mr Sheffield adopted was to work with me at the bench and occasionally to take me into his surgery, where I saw him operate. I was also at his side when he saw the gratis patients who came to him each morning at nine o'clock to be relieved of pain. I did all his mechanical work during my apprenticeship, and after about three months he gave the care of the gratis patients into my hands, and I well remember when I went down alone to extract my first tooth, which I am happy to say I accomplished successfully. From that time I selected from among the cleanest of these poor people those who needed teeth stopped, and attended to them. Mr

Sheffield used to examine the teeth when I had prepared them, and afterwards when the filling was completed, but he soon left me to my own devices, except when any great difficulty arose, when he always came to my help. After about a year or so, he left me in charge of his private patients when he went for his holiday . . .

However, during the early years, probably no more than half entered the profession by apprenticeship. There was, after all, no legal requirement for a 'dentist' to have any training; as an observer remarked in the 1840s, 'a brass plate and brazen effrontery [were] all the diplomas necessary'.[10]

The state of the profession in the mid-nineteenth century
By the middle of the nineteenth century, the number of dentists in the country had grown at a remarkable rate. In 1850 there were about 300 in London and nearly 400 in the provinces. They were also to be found in Scotland (mostly around Glasgow and Edinburgh) but resident dentists were singularly lacking in most of Wales. Permanent practitioners made their appearance in Ireland at the beginning of the nineteenth century. Within the profession was a very small number of medically qualified men who had turned to dentistry and a larger group of chemists and druggists who also practised as dentists.

Many of the dentists now in practice were members of dental families, who had had no choice but to go into the family business. (A few of these 'dental dynasties' are still with us today, having carried on a tradition of practising dentistry for anything up to 150 years.) Alongside these career dentists were the still-growing numbers of opportunists, who picked up a few tricks of the trade and set out to try dentistry for themselves. These were the ones who frequently abandoned it again quite quickly.

For those who persisted, the rewards could be substantial, a fact presumably known to those looking around for a new line of business. The available evidence suggests that many dentists could expect to gross a good £700–800 pa, with higher returns for the fashionable London dentist (fig. 3.5). This was far in excess of the income to be expected from general medical practice (probably £200 in the provinces in the 1840s and £400 in London).

Some of these high incomes were undoubtedly achieved by less

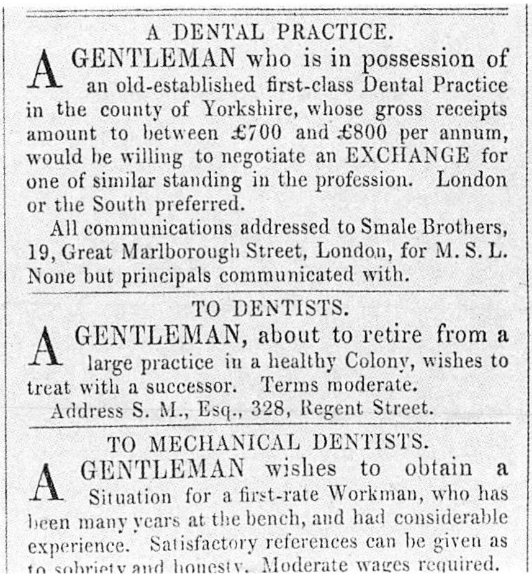

A DENTAL PRACTICE.

A GENTLEMAN who is in possession of an old-established first-class Dental Practice in the county of Yorkshire, whose gross receipts amount to between £700 and £800 per annum, would be willing to negotiate an EXCHANGE for one of similar standing in the profession. London or the South preferred.

All communications addressed to Smale Brothers, 19, Great Marlborough Street, London, for M. S. L. None but principals communicated with.

TO DENTISTS.

A GENTLEMAN, about to retire from a large practice in a healthy Colony, wishes to treat with a successor. Terms moderate.

Address S. M., Esq., 328, Regent Street.

TO MECHANICAL DENTISTS.

A GENTLEMAN wishes to obtain a Situation for a first-rate Workman, who has been many years at the bench, and had considerable experience. Satisfactory references can be given as to sobriety and honesty. Moderate wages required.

Fig. 3.5 Trading in practices (*British Journal of Dental Science* 1856; 1: 197).

than ethical means since, in the absence of any legal or professional control over the practice of dentistry and no way for the potential patient to distinguish the trained practitioner from the quack, the door was wide open for sharp practice and sheer incompetence. No one was more scathing or vituperative about the state of dentistry in the middle of the nineteenth century than dentists themselves. In article after article they exposed the unscrupulous and bemoaned the straits in which dentistry found itself. Nevertheless, when James Robinson, a prominent London dentist, tried in 1842 to form a dental society to combat these ills, he received so little support that the idea had to be abandoned. The truth was that the 'respectable' dentists did not want to associate with their less reputable confrères, but no one could decide where the line should be drawn between the two groups.

Reform of the profession, 1856–1956

Out of this inertia eventually emerged two rival factions dedicated to furthering the cause of dental education and professionalism. The first, the College of Dentists, arising out of an open letter to the *Lancet* in 1855 from a young Croydon dentist, Samuel Lee Rymer,[11] stood for an independent profession. The second, the Odontological Society of London, was begun by a number of influential London dentists (including John Tomes), some of whom had already tried, 12 years earlier, to interest the Royal College of Surgeons of England (RCS) in setting up a dental diploma. Both societies held their first meeting in November 1856, on consecutive days, the Odontological Society winning the race, by a political manoeuvre, to be counted the first professional dental society in England.

The RCS was finally enabled to grant a Licence in Dental Surgery (LDS) in 1858. By the time of the first examination, in March 1860, the Odontological Society had set up its own Dental Hospital of London and London School of Dental Surgery (until 1985 the Royal Dental Hospital) to train students (fig. 3.6). Undeterred, the College of Dentists set up the Metropolitan School of Dental Science in 1859 and the National Dental Hospital in 1861 to train students for its own examination. (They became the University College Hospital Dental School in 1914.) Only in 1863 did the College of Dentists finally capitulate and abandon its stand for independence. The two rivals then merged to form the Odontological Society of Great Britain (now the Odontological Section of the Royal Society of Medicine).

Few dentists obtained the new LDS qualification, partly because the only hospitals and schools recognised by the Royal College of Surgeons were in London. Renewed agitation for reform in the 1870s finally let to the Dentists Act of 1878 and, what appeared at the time

Fig. 3.6   The Royal Dental Hospital and School in Soho Square, London, 1937 (BDA Museum).

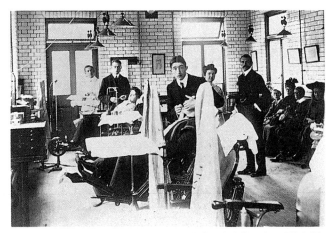

Fig. 3.7 Students at work in the National Dental Hospital, London (BDA Museum).

to be the answer to all problems, the first *Dentists Register* of 1879, administered by the General Medical Council. Now, at last, only those who were on the register could call themselves 'dentist' or 'dental surgeon'. After an initial amnesty for all those who could show they had been in bona fide practice before the passing of the Act, new entrants must have obtained the LDS. The Royal Colleges of Surgeons of Edinburgh and Glasgow and in Ireland were empowered to grant their own Licences. Dental Schools were set up outside London, in some cases attached to existing dental hospitals or dispensaries (as in Birmingham and Liverpool) to prepare students for the new licences (fig. 3.7).

However, all was not well. Intense lobbying of members of Parliament (sometimes while they were undergoing treatment in the dental chair) and an extensive campaign had been required in 1878 to achieve the passage of the Dentists Bill in the face of determined opposition from members of the medical profession. Provisions to prevent irregular initial registration and other safeguards were excised before it became the Dentists Act 1878. The new British Dental Association (formed in 1880) found itself much occupied with bringing prosecutions against those practising illegally, since the General Medical Council seemed reluctant to exercise its duties in this direction.

Fig. 3.8   How to practise dentistry without being on the Register.

Loopholes were found in the Act and exploited from the first. There being no requirement to register, many did not do so and, provided they did not use titles protected by the Act, were able to practise without professional education or ethical responsibility (fig. 3.8). It was also discovered that a person could, by paying £10 to register himself and six others (such as a spouse and servants) as directors, form a dental company totally exempt from all sanctions under the Dentists Act.

Unregistered practice and blatant exploitation by ignorant practitioners became so widespread that pressure of public opinion led to investigation by a committee set up by the Privy Council. This issued, in 1919, a horrifying report on the state of dentistry at that time. The resultant Dentists Act 1921 restricted practice to registered persons; again, as in 1879, after an initial amnesty, only qualified dentists could join the Register. There was set up a board of 13 members and a chairman (appointed by the Privy Council) which depended on the General Medical Council for approval of all its recommendations. This new Dental Board of the United Kingdom kept *The Dentists Register*. In the course of its existence it applied over half a million pounds, from the fees of dentists for retention on the register, to the improvement of dental education, bursaries for students, dental research and dental health education.

The profession finally achieved self-government in 1956 with the setting up of the General Dental Council, which took over the functions of the Dental Board.

Dental education in the twentieth century

In 1900 the first university degree in dental surgery was established in Birmingham and first awarded in 1901; other universities soon followed suit. From 1948, all the dental schools were attached to universities and were thereafter funded by the University Grants Committee (now the University Funding Committee).

Higher Dental Diplomas were established by the Scottish Royal Colleges in 1920 and the RCS followed with its Fellowship in Dental Surgery in 1947. Diplomas in such specialties as orthodontics, dental public health and general dental practice followed. Higher dental degrees date from the beginning of the century and multiplied after the closer connection of dental education with the universities after 1948.

Ancillary staff

Shortages of manpower, particularly in the first half of the century, led to the introduction of dental ancillaries. Powers were granted in the Dentists Act 1921 to employ 'dental dressers', after only a few weeks' training, to treat school children. Some Midland local authorities availed themselves of this section of the Act until 1925.

The Dentists Act 1956 enabled the General Dental Council to establish a school for dental auxiliaries at New Cross General Hospital. After registration they were entitled to perform a limited number of procedures under direct supervision in the public dental service. Their title was later changed to dental therapist, the school closed and their training was moved to the London Dental Hospital.

In Great Britain, dental hygienists were first trained and used in the WAAF in 1943. Granted certification by the Ministry of Health in 1949, supervision of their training at dental hospitals and their subsequent registration eventually became the responsibility of the General Dental Council. Dental hygienists are able to scale and polish

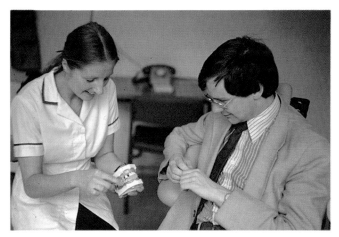

Fig. 3.9   A hygienist gives oral hygiene instruction to a patient.

teeth, give oral hygiene instruction and apply preventive agents to the teeth, to the prescription of a dentist (fig. 3.9).

At the time of the Dentists Act 1878 and for many years after, the staff of dental schools were general dental practitioners, many serving unpaid. Dental research was undertaken entirely by talented general dental practitioners in their leisure time and at their own expense. A century later, dentists are not only in general practice but hold academic posts in their specialty, provide hospital dental services (many as consultants), staff a community dental service, are employed in industry and serve within the administration of the Department of Health.

4

The Patients

The history of dentistry in this country shows treatment being made available to an increasing number of people, because more came to need and want its services and more could afford them.

Poorer people

In pre-industrial days, poorer people, especially in rural populations, probably had less experience of caries than their masters, who could afford to indulge their taste for sugar.

It was the Industrial Revolution, with its improved transport, which first brought the life of the towns to the country and, at a later stage, took hordes of people to the expanding towns. More people came into contact with a diet high in refined foods. When import duty was reduced on sugar, consumption rose rapidly (from 19 lb per head per year in 1859 to 90 lb in 1900), and so did the level of caries in the poorer sections of society. Extraction was often the only remedy for those without means, but still only as a last resort, when the pain became intolerable, since there were no local or general anaesthetics to eliminate pain before the mid-nineteenth century (fig. 4.1).

Such people were the target consumers of the many toothache cures available or the products promising to improve the appearance of the teeth. In 1776, a Leicester dentist was selling a liquid which, when painted on the teeth, made them 'white as alabaster'.[12] There were many eighteenth century advertisements for dentifrices and toothpowders, such as Jackson's 'true British powder for the teeth'. Teething powders were probably much in demand. 'Teeth' featured prominently in the weekly bills of mortality (summaries of the numbers and causes of death), although it was probably not teething itself which carried off the unfortunate infants so much as fever.

Even if the poorer groups in society had wanted to have artificial teeth made for them, the fees would have made it impossible for

Fig. 4.1 Extraction without benefit of anaesthesia, depicted by De Cari, 1820 (R. A. Cohen's collection).

them. Early dentures were carved from solid ivory, a highly skilled, time-consuming and expensive job. Artificial teeth became available to the mass of the population only after vulcanite came into more common use in 1881 (when the original patent expired) and the price dropped to about £5, a week's wage for an agricultural worker of the day. With time, raised standards of living enabled more people to be able to consult a dentist for treatment if they wished, but dentistry was still closed to the majority of the population. Fortunately, some dentists set aside time to treat the poor without charge.

Dental dispensaries
In the mid-nineteenth century a few practitioners established dental dispensaries. They gave their services free, other costs being met by the benevolence of rich people. Apart from providing poor people with dental care previously available only to the wealthy, such institutions also enabled students to learn the art of dental practice. In 1839 the London Institution for the Diseases of the Teeth opened its doors in Windmill Street. In the first four years it gave relief to 5903 people. In 1855 the London Dental Dispensary opened near Regent's Park. Similar dispensaries were founded in Birmingham (1858), Edinburgh (1860) and Liverpool (1860) which have survived until the present day as the dental hospitals in those cities; a large number of

others which were founded in the nineteenth century have since ceased to function under their original names, having been transferred to the National Health Service in 1948.

School dentistry

A lecture to the British Dental Association in 1885 highlighted the appalling condition of children's teeth. Following a number of further reports (fig. 4.2), the British Dental Association declared that pain and sepsis in the lower social classes was a public scandal. It pressed for a school dental service to be provided and paid for by the state, including compulsory dental inspections and health education.

Some Poor Law and Public schools had already appointed dentists. These men founded the School Dentists Society in 1896. Its major aim was to inform education authorities about the importance of prevention rather than only of treatment. And it fought for a service for *all* children, not just those *seeking* care.

The need for school services received a major impetus during the Boer War when half the adult males were found to be unfit for

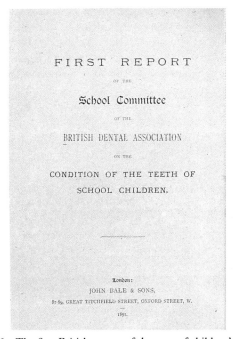

Fig. 4.2 The first British survey of the state of children's teeth.

military service, many for dental reasons. Of 69 553 men inspected, 4400 were rejected on account of 'loss or decay of many teeth'. Even those men who got through the net still had problems. Soldiers of the Cheshire Regiment suffered so much from gastric troubles after eating badly chewed food, that mincing machines had to be supplied. Clearly, something had to be done to improve the nation's teeth.

The 1907 Education Act made it compulsory for education authorities to arrange dental inspections, but not treatment. However, a Cambridge dentist, George Cunningham, had already gone a step further. In 1907, he had been responsible for opening the Cambridge Dental Institute, the first UK children's clinic (fig. 4.3). It was endowed by a Cambridge don, convinced by Cunningham of its value. Not surprisingly, Cunningham is regarded as the father of the British school dental service.

Education authorities around the country eventually followed his lead and paid for the establishment and staffing of school dental clinics. In addition, there were many preventive programmes, such as the Toothbrush Clubs in London schools, where children learned to care for their teeth (fig. 4.4). Brushes were sold cheaply. It was not unusual for children to buy one as a birthday present for a parent, paying for it in instalments. Under the 1944 Education Act, local education authorities were obliged to secure treatment for all schoolchildren.

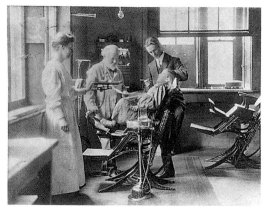

Fig. 4.3   George Cunningham (centre) in his Cambridge clinic (BDA Museum).

Fig. 4.4 The Shepperton Toothbrush Club, 1912.

Maternity services

From 1918, local authorities had the power, but not the duty, to provide care for expectant and nursing mothers and preschool children. By 1944, 407 local authorities had made such arrangements. Even where facilities existed, advantage was seldom taken of them.

General dental services

The 1911 National Insurance Act allowed Approved Societies to provide dental treatment if they had surplus funds. Such surpluses became available in 1922, in which year the first dental benefit subsidised by the state was paid. It consisted of full or part payment (not less than 50%) of the cost of approved treatment. A dentist was under no obligation to provide treatment for insured persons but, if he did so, he had to provide all the treatment necessary to render the patient dentally fit.

Treatment was limited to workers earning less than a certain amount, with a period of membership of usually two-and-a-half years; it did not extend to their relatives. As it was not available before the age of 17, children leaving school at the age of 14 had no assistance in obtaining dental treatment for these three important years.

Dental disease in the general population was serious and widespread, with a lamentable failure among most people to appreciate its importance. Demand for even the available services was very low and often sought too late; the only treatment possible was frequently wholesale extraction of teeth and the provision of dentures.

The situation appeared little better 20 years later. By 1943 some 5000 Approved Societies provided dental benefits for nearly 14 million members, 75% of the insured population. Even then, mostly because of the cost, only 6% of the eligible people were treated each year. It was the common experience of insurance dentists that people did not resort to treatment until the teeth were unsavable and there was severe oral sepsis.

Some free or low-cost treatment was available to the non-insured at the voluntary dental hospitals and in a limited form at a number of general hospitals. Some local authorities made provision for the dental treatment of certain categories of patients, such as those with tuberculosis; others even provided treatment at a small charge for the general population in municipal dental clinics. Some large industrial firms provided care for their own employees. However, most non-insured people had to resort to private treatment, which generally ruled out anything but emergency measures.

Lack of means was not the only restraint. An interdepartmental committee on dentistry reported in 1944 that the public was ill-educated and apathetic in regard to the care of the teeth: 'This attitude springs mainly from a national fear of pain and lack of any real understanding of the importance of dental health . . . The state of the dental health of our population is bad and its effect on their general health is bad'. Amongst recruits to the army during World War II, 90% of the men required dental treatment, 13% already possessed dentures and a further 10% needed them. As regards children, only one in ten of 10 000 Scottish five-year-olds were free of tooth decay; 70 000 (35%) of their teeth were decayed or missing. One Approved Society found that 12% of all young people under 19 applying for treatment needed full upper and lower dentures—they had no teeth left of their own.

Not surprisingly, faced with the ever-mounting evidence of the low level of dental health of the nation, when the Government began to consider the possibility of a National Health Service, dentistry was included in its plans. William Beveridge, in his 1942 *Report on Social Insurance and Allied Services*, maintained that there was 'a general demand that dental services should become statutory benefits available to all under health insurance'. He also pointed out that there was a need for a 'change of popular habit from aversion to visiting the dentist until pain compels, into a readiness to visit and be inspected periodically'.

When the NHS came into being in 1948, dental care was included in its provisions. The demand for treatment was immediately overwhelming and took all parties concerned by surprise. A government which had introduced a free health service found itself imposing charges to control expenditure.

Some form of charge for treatment has remained ever since for most groups of people. Nevertheless, the inclusion of dentistry in the NHS has had a radical effect on the state of the nation's dental health. For those under the age of 50, in particular, looking after the teeth and seeking regular professional advice has become a normal part of the experience of all sectors of society, not just of a fortunate élite.

5

Dental Literature

Early sources

Teeth, their diseases and remedies for treating toothache were described as long ago as 5000 BC on a Sumerian clay tablet found in Mesopotamia and in Chinese, Egyptian and Roman sources. Medical texts would often include sections on the teeth, but it was not until the late sixteenth century that the first book devoted entirely to dentistry appeared, the *Artzney Buchlein*, published anonymously in Germany in 1530 (fig. 5.1); a small booklet of 44 pages, it was intended for the public rather than the medical profession and discussed such topics as toothache, decay, loose teeth and worms in the teeth. Fourteen editions were published, some entitled *Zene Artzney*, the last in 1576.

In 1563 Bartholomew Eustachius published the first dental book for the medical and dental profession. His *Libellus de dentibus* was a treatise on the anatomy and histology of the teeth.

By the end of the sixteenth century over forty dental books had been published in Germany, France, Spain and Italy. The first book in English appeared in 1685, Charles Allen's *The operator for the teeth*, a booklet of 56 pages printed in York (fig. 5.2). Like the *Zene Artzney*, it was intended for the general public.

Eighteenth and nineteenth century books

A milestone in the advancement of dentistry was the publication of Pierre Fauchard's *Le chirurgien dentiste* in 1728. French dentists had preferred to guard their 'professional secrets', declining to publish results of experiments or details of their methods. Fauchard, by contrast, wrote down all he knew about contemporary dental science and practice and described his own inventions, with the intention of giving his peers and successors a scientific reference book. Dental historians therefore have an invaluable source on eighteenth century

Fig. 5.1   *Artzney Buchlein*, 1530, published anonymously.

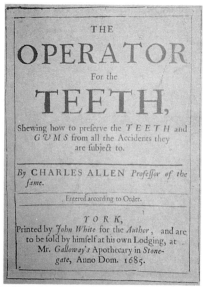

Fig. 5.2   *The operator for the teeth* by Charles Allen, 1685.

dentistry. Fauchard stressed the importance of sound teeth, described appliances to straighten crooked ones, and the manufacture of dentures. The second edition of *Le chirurgien dentiste* (fig. 5.3) was translated into English by Lilian Lindsay in 1946.

In the 1750s and 1760s dental literature began to develop, an important early text being Thomas Berdmore's *A treatise on the disorders and deformities of the teeth and gums*, published in 1768. This covered the whole field of dentistry and was a popular book, running into three editions, one reprinted as late as 1844. In common with others, this book was translated into other languages.

John Hunter, the surgeon and anatomist, gave a scientific basis to dentistry with his *The natural history of the human teeth*, published in 1771 and including numerous meticulous illustrations. The same title was used by Joseph Fox in 1803 for his classic textbook, which was based on the lectures given at Guy's Hospital.

In 1801 R. C. Skinner produced the first American dental book, *A treatise on the human teeth*, intended for the layman. The first American dental textbook, *The dental art: a practical treatise on dental*

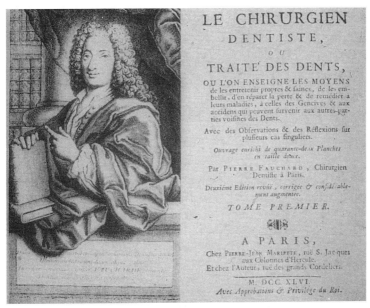

Fig. 5.3 *Le chirurgien dentiste* by Pierre Fauchard, second edition 1746.

surgery, was written in 1839 by C. A. Harris. This was one of the most widely used books on the subject; the thirteenth edition appeared in 1897 and was reprinted for the last time in 1912. In England, *A system of dental surgery* by Sir John Tomes, published in 1859, in its various editions remained the standard textbook for fifty years. Tomes is rightly regarded as the father of British dentistry.

By the 1880s dental specialties were beginning to emerge; reflecting this trend, books on different aspects of the subject, such as materials, crown and bridge work, denture construction and orthodontics, became the norm, replacing general compendia. J. E. Garretson had written the first oral surgery textbook, *Treatise on diseases and surgery of the mouth, jaws and associated parts* in 1869 and in 1880 N. W. Kingsley wrote on the scientific treatment of irregularities of the teeth in *A treatise on oral deformities*. Of especial significance is W. D. Miller's *Microorganisms of the human mouth*, published in 1890, in which he expounded the chemico-parasitic theory of the cause of caries, still broadly accepted today.

E. H. Angle's *Notes on orthodontia* (1887) was expanded to the definitive seventh edition, published in 1907 as *Treatment of malocclusion of the teeth*. G. V. Black produced his magnum opus *A work on operative dentistry* in 1908; the seventh edition of this classic work was dated 1948, and ran to four volumes.

It is not feasible here to single out individual books from the hundreds published during the twentieth century. The majority are specialised textbooks and monographs for dental students, practising dentists and research workers. Advances in printing techniques have allowed more illustrations to be included; the 'colour atlas', comprising photographs showing techniques and diseases, has become increasingly popular.

Reference books

The first dental dictionary was C. A. Harris's *A dictionary of dental science, biography, bibliography and medical terminology*, first issued in 1849, with its sixth edition in 1898. An English, as opposed to American, dictionary did not appear until 1962: *Heinemann modern dictionary for dental students* by J. E. H. Fowler (later published under her married name of Fairpo).

The Dentists Register, a list of those registered to practise in Britain, started its annual publication in 1879; it gives the name and address of each dentist, qualifications held and when and where they were obtained.

The development of dental journals

As the numbers of dentists increased and the profession became more organised, the need for regular and up-to-date information became apparent. Periodicals produced at frequent intervals could keep readers abreast of news on the scientific, clinical and political fronts and, via correspondence columns, provide a forum for the exchange of views. Advertisements would describe new products and equipment and enable practitioners to advertise their own services, as they were quite free to do before registration was introduced. Journals therefore developed to fulfil the dual functions of recorders and newspapers.

The producers of journals in the nineteenth century fell into four main categories: individual dentists or groups of dentists, societies, dental manufacturers or traders and, to a lesser degree, dental schools.

The first periodical to be founded was the *American Journal of Dental Science* (1839–1909), begun as the organ of the American Society of Dental Surgeons. The first two British attempts in this field, both made by James Robinson, were short-lived. The *British Quarterly Journal of Dental Science* had only two issues, in 1843, and *The Forceps*, launched in 1844, died the following year after 41 issues.

The third attempt resulted in an authoritative and successful journal, the *British Journal of Dental Science* (1856–1935). This was acquired at the end of its first year by some supporters of the Odontological Society of London. Adherents of the rival College of Dentists brought out their own *Quarterly Journal of Dental Science* (1857–1859), an outspoken publication replaced by the more diplomatic *Dental Review* (1859–1867). The Odontological Society of London (later Great Britain) published its own *Transactions* from 1856 to 1907, when the Society became part of the Royal Society of Medicine.

The *Monthly Review of Dental Surgery*, begun in 1872, was acquired in 1880 by the newly formed British Dental Association. Since 1903 it has been entitled the *British Dental Journal*.

The first British journal to emanate from the dental trade was the *Dental Record*, produced by the Dental Manufacturing Company

Ltd. This was the sector which published the most influential journals in the United States, the most prestigious of which was *Dental Cosmos* from S. S. White Company. Other countries were producing journals in their own languages. Germany, for example, has a longstanding tradition of periodical publication dating back to 1860.

Finance was a perennial problem and the main reason for the failure of journals. Production of any periodical was costly, but sales of independent and society publications could be severely affected by the lower subscription price of proprietary journals which were subsidised by their parent companies.

Twentieth century journals

As the decades passed, it became economically impossible for individuals to own journals; their place was taken by commercial publishers. Membership of national and local dental societies was on the increase and these bodies started to issue regular publications, many starting as newsletters or bulletins and gradually evolving into formal, often prestigious journals. The National Dental Association, for example, launched its *Official Bulletin* in 1913; this later developed into the *Journal of the American Dental Association* in 1922. The independent stance offered by the society publications came to be preferred to the proprietary journals whose impartiality was in question.

The evolution of dental specialties was reflected in the development of specialist titles, notably in the 1930s and particularly in the United States. Although the earliest specialist journals were generally of American origin, an important exception to this was the *Transactions of the British Society for the Study of Orthodontics*, first issued in 1908.

The 1960s and 1970s were growth years for specialist journals, with new titles in such fields as endodontics, children's dentistry, prosthetics and periodontology, from societies and commercial publishers.

Identification of journal articles on particular subjects or by specific authors is made possible by the *Index to dental literature*. This vital tool indexes major English language dental journals back to their beginnings, in 1839.

Literature on the history of dentistry

Excellent, scholarly texts, but possibly difficult to obtain, are

V. Guerini's *A history of dentistry from the most ancient times until the end of the 18th century* (1909 and reprinted in 1967 and 1968) and W. Hoffmann-Axthelm's *History of dentistry* (1981).

More easily available, for instance in public libraries, is M. E. Ring's *Dentistry: an illustrated history* (1985). This large book is profusely illustrated and provides an eminently readable account of dental history. *The advance of the dental profession* (1979) charts the first hundred years of the British Dental Association. The Lindsay Society (previously the Lindsay Club) is the historical society of the British Dental Association and publishes *Dental Historian* annually; issues are available from the Librarian, British Dental Association, 64 Wimpole St, London W1M 8AL, from whom further information about this society can be obtained.

Appendices

Appendix 1   References cited in the text

1 Miller W D. *Microorganisms of the human mouth.* Philadelphia, 1890.
2 Verney F P. *Memoirs of the Verney family* 1892; **2**: 235. 1894; **3**: 39. *Cited by* Campbell J M. *Dentistry then and now.* 3rd edition, p 291. Privately printed, 1981.
3 Hoffman-Axthelm W. *History of dentistry.* Translated by Koehler H M. p 121. Chicago: Quintessence, 1981.
4 Dickenson F. *A pretious treasury or a new dispensary,* book 2. London, 1649.
5 Martial. *Epigrams,* Book I, No. 72 (first century AD).
6 Fox J. *Natural history of the human teeth.* London: Cox, 1803.
7 Angle E H. Classification of malocclusion. *Dental Cosmos* 1899; **41**: 248–264, 350–357.
8 *Postman,* January 30, 1696.
9 Woodhouse A J. Reminiscences of fifty-four years in the dental profession. *Journal of the British Dental Association* 1897; **18**: 21–31.
10 Robinson J. *Forceps* 1844–45; **1**: 62.
11 Rymer S L. Necessity for a college of dental surgery. *Lancet* 1855; **2**: 181 (letter).
12 *Leicester and Nottingham Journal* December 21 1776.

Appendix 2 Chronology

c.5000 BC	A Sumerian clay tablet described worms as the cause of caries and gave remedies for toothache
4000–3000 BC	Agurreda, a Sanscrit record, described oral anatomy, diseases of the mouth and Hindu oral hygiene methods
1500 BC	The Papyrus Ebers included tooth diseases and remedies
600–300 BC	The Etruscans made dental appliances and primitive bridge work
450 BC–AD 100	Dentistry and teeth are mentioned in Roman medical books (Celsus, Pliny and Scribonius) and by the poets Martial and Juvenal
1530	The first book solely on dentistry, *Artzney Buchlein*
1563	Bartholomew Eustachius published the first book on dental anatomy, *Libellus de dentibus*
1683	Antony van Leeuwenhoek identified oral bacteria using a microscope
1685	The first dental book in English, Charles Allen's *The operator for the teeth*
1728	Pierre Fauchard's *Le chirurgien dentiste*. (Second edition in 1746)
1764	James Rae gave the first course of lectures on the teeth at the Royal College of Surgeons in Edinburgh
1771	John Hunter's *The natural history of the human teeth* gave a scientific basis to dental anatomy
1778	John Hunter's *Practical treatise on the diseases of the teeth*
1780	William Addis of Clerkenwell produced the first toothbrush of modern design
1788	Nicolas Dubois de Chémant described porcelain dentures in his *Dissertation sur les avantages des nouvelles dents et rateliers artificielles*, published in Paris
1799	Joseph Fox began to give lectures on the teeth at Guy's Hospital, published in enlarged form in 1803 and 1806 as *Natural history of the human teeth*
1820	Claudius Ash established his dental manufacturing company in London

1831	James Snell designed the first reclining dental chair
1839	The first dental school, journal and society founded: Baltimore School of Dental Surgery, *American Journal of Dental Science* and American Society of Dental Surgeons
	London Institution for the Diseases of the Teeth opened
1830s–1890s	The 'amalgam war' waged over the use of amalgam as a filling material
1844	Horace Wells demonstrated nitrous oxide as an anaesthetic
1846	W. T. G. Morton demonstrated ether as an anaesthetic
1851	Vulcanite, to be used for denture bases, invented by Charles Goodyear
1856	Formation of the Odontological Society of London and of the College of Dentists
1858	Birmingham Dental Dispensary opened (the oldest now surviving)
	Opening of the Dental Hospital of London, the first establishment in Britain for the clinical training of dentists
1859	Opening of the first dental schools in Britain, the London School of Dental Surgery and the Metropolitan School of Dental Science
1860	Royal College of Surgeons of England awarded the first Licences in Dental Surgery
1863	Odontological Society of London and the College of Dentists merged to form the Odontological Society of Great Britain
1866	Lucy B. Hobbs became the first woman to qualify as a dentist in Ohio, USA
1867	Leber and Rottenstein implicated bacteria in caries
1871	James Beall Morrison invents foot treadle engine
	Introduction of silicate cement, a tooth-coloured filling material
1878	First British Dentists Act: titles were protected but registration not compulsory

1879	First Register
1880	British Dental Association founded
1884	Cocaine introduced as a local anaesthetic by Carl Koller
1890	W. D. Miller formulated his 'chemico-parasitic' theory of caries in *Microorganisms of the human mouth*
1895	Lilian Lindsay (née Murray), having been refused entry to the London schools on the grounds of her sex, graduated from Edinburgh as the first woman to qualify as a dentist in Britain
	Wilhelm Conrad Röntgen discovered x-rays; speedy application to dentistry
1900	Fédération Dentaire Internationale founded
1901	First British dental degree awarded by the University of Birmingham
1905	Alfred Einhorn introduced Novocaine (procaine) as a local anaesthetic
1907	William H. Taggart demonstrated a casting method to produce gold inlays
	Odontological Society became a Section of the Royal Society of Medicine
	George Cunningham organised the Cambridge Schoolchildren's Dental Institute
1921	Dentists Act: only registered practitioners allowed to practise
c1932	Synthetic resins introduced for denture bases
1943	First dental hygienists were trained by the Women's Auxiliary Air Force
1945	Grand Rapids, USA, introduced a water fluoridation scheme
1948	Inception of the National Health Service
1955	Michael Buonocore described the acid-etch technique
1956	Dentists Act; General Dental Council set up
1957	High-speed turbine handpiece introduced
1959	Fluoride toothpaste marketed in Britain
1962	First dental auxiliaries qualified at New Cross General Hospital

1968	First national dental survey of the adult population in England and Wales
1984	Dentists Act; constitution and powers of the General Dental Council were modified
1985	White filling material for posterior teeth marketed

Appendix 3 Glossary

Amalgam An alloy of mercury with silver and/or other metals, used to fill teeth

Anaesthesia Loss of sensation. In *general* anaesthesia consciousness is lost but in *local* anaesthesia only a small area is affected. In *intravenous* anaesthesia, the drug is given through a vein

Analgesics Drugs which prevent pain from being felt

Annealed Made more flexible by heat treatment

Approved Societies Friendly Societies authorised by the Ministry of Health to provide medical and other benefits to their members

Articulator A hinged device upon which plaster-of-Paris casts of the jaws can be mounted. Used during the construction of dentures to ensure their correct occlusion (*see below*)

Astringents Drugs which reduce bleeding by making the blood vessels shrink

Bridge One or more artificial teeth permanently attached to the adjacent natural teeth

Bur A small drill inserted in a dental handpiece

Calculus Hardened plaque, tartar

Cauterise To destroy tissue by burning with a hot iron or caustic substance

Cohesive gold Small pellets of gold which are hammered together to form a solid mass

Crown The visible part of a tooth; gold, porcelain or acrylic substitute for natural crown fitted over the prepared tooth or root

Demineralise To remove the hard part of a tooth by the action of acids

Dentine The hard tissue under the enamel which surrounds the pulp and forms the bulk of a tooth

Denture base Plates made to fit the mouth, to which artificial teeth are fastened

Enamel The hard, white outer covering of the crowns of the teeth

Endodontics Treatment of the diseases of the pulp and root canals of the teeth

Fluoride A salt of fluorine which, when added to the water supply or applied to the teeth by toothpaste or other means, protects against tooth decay

Formalin A solution of formaldehyde used as a disinfectant

Gutta-percha A white, rubbery material with various uses in dentistry, such as filling root canals

Inclined plane An orthodontic appliance having a surface which
when bitten upon, applies a pressure that causes a tooth to move

Inlay A filling, usually cast in gold; prepared in the laboratory to fit
a model of the cavity before being cemented into the tooth

Iodoform An antiseptic based on iodine

Labial arch An orthodontic appliance with springy wire which fits
around the outer surfaces of the teeth to bring them into the correct
position

Lance To pierce an abscess with a sharp instrument to drain off pus

Ligatures Threads or wires used, for example, to tie teeth together

Malocclusion Abnormal closure of the jaws caused by irregular teeth

Mastication Chewing

Narcotics Drugs which produce drowsiness and relieve pain

Obturator An appliance used to replace missing tissue; particularly
a device to cover a hole in the palate

Occlusion The way in which the teeth meet together on biting and
chewing

Organic Any material of animal or vegetable origin

Orthodontics Treatment to prevent or correct irregularities of the teeth

Palate Roof of the mouth

Periodontal disease Disease of the gums and tissues supporting the
teeth

Pharmacopoeia A list of drugs with their properties and uses

Pinchbeck An alloy of copper and zinc, imitating gold

Plaque The deposit on teeth, formed mostly of bacteria

Platina An alloy of platinum and several other metals

Poor Law schools Schools attached to workhouses

Posterior Towards the back (of the mouth, face, etc)

Prosthetic dentistry Replacement of wholly or partially missing teeth
by artificial substitutes such as dentures, crowns or bridges

Pulp The soft, innermost part of the tooth, containing the nerves
and blood vessels

Resurrectionists Body-snatchers

Retention The ability of a denture to stay in place despite the force
of gravity, sticky foods, etc.

Root canals The passages in the roots of teeth where nerves and
blood vessels enter the dental pulp

Scaling Freeing the tooth surface of calculus by the use of special
instruments

Sepsis   The presence of harmful bacteria in the body

Shell crowns   Metal coverings for teeth

Silicate   A white material for filling teeth

Statutory Benefits   Payment during sickness, unemployment, etc, authorised by law

Swage   To shape metal by hammering it on to a metal model

Tartar   Calcified plaque

Unerupted teeth   Teeth which have not yet become visible in the mouth, although present in the jaw

Vulcanite   A hard material produced by heating rubber with sulphur at high temperature and pressure

Index

Acid-etch technique 65
Acrylic dentures 33
Addis, William 63
Agurreda 63
Allen, Charles 55, 63
Amalgam 24, 32, 64, 67
American Journal of Dental Science 59, 64
American Society of Dental Surgeons 59, 64
Anaesthesia 20, 67
  general 27–31, 34
  local 33–34
Ancillary staff 45–46, 65
Angle, E. H. 25–27, 58
Anodyne pills 12–13
Apprenticeship 38–40
Articulators 30, 67
Artzney Buchlein 55, 56, 63
Ash, Claudius 18, 63

Baltimore School of Dental Surgery 64
Barbers 35
Benefits, dental 51, 52
Berdmore, Thomas 57
Beveridge, William 52
Birmingham Dental Dispensary 64
Black, G. V. 58
Blair, James 38
British Dental Association 43, 49, 59, 61, 65
British Dental Journal 59
British Journal of Dental Science 59
British Quarterly Journal of Dental Science 59
Buonocore, Michael 65

Cabinets, dental 30
Cambridge Schoolchildrens' Dental Institute 50, 65
Caries, dental 31, 34, 47, 63, 64
  chemico-parasitic theory 3, 58, 65

Chair, operating 30, 64
Charges, treatment 53
Chemico-parasitic theory 3, 58, 65
Chirurgien dentiste, Le 55, 57, 63
Chloroform 28
Cleaning, tooth 5–7
Clover, Joseph 30
Cloves, oil of 9
Cocaine 33, 65
College of Dentists 41, 42, 59, 64
Composite materials 32
Crowns 21–22, 24, 58, 67
Cunningham, George 50, 65

Davy, Humphrey 27
De la Roche, Peter 36
Dental Board of the United Kingdom 45
Dental Cosmos 60
Dental Hospital of London 64
Dental Manufacturing Company Ltd 59
Dental Record 59
Dental Review 59
Dentists 37–41
  apprenticeship 38–40
  incomes 40–41
  professional reforms 41–45
  travelling 37, 38
Dentists Act (1878) 42, 43, 44, 46, 64
Dentists Act (1921) 44, 45, 65
Dentists Act (1956) 45, 65
Dentists Act (1984) 66
Dentists Register 43, 44–45, 58–59, 64
Denture bases 19–20, 67
Dentures 10, 11–12, 15–16, 47–48, 52, 58
  all-in-one blocks 18, 19
  decay of 16
  gold 19–20, 36–37
  human replacement teeth 15, 16
  impression-taking 15, 20–21
  ivory 11–12, 15, 16, 20, 48

Dentures—*contd*
 partial 16, 19
 porcelain 16–20, 63
 springs 15–16
 synthetic resin 32–33, 65
 terro-metallic teeth 18, 19
 tube teeth 18, 19
 vulcanite 20, 32, 48
Dispensaries, dental 48–49
Dressers, dental 45
Dubois de Chémant, Nicolas 63

Education Act (1907) 50
Education Act (1944) 50
Egypt, ancient 1–2, 9, 10, 63
Einhorn, Alfred 65
Engine, dental 24, 30, 64
Ether 28, 64
Etruscans 11, 63
Eustachius, Bartholomew 55, 63
Evans, Thomas W. 30
Extraction 9–10, 35, 47, 52
 instruments 13–14

Fauchard, Pierre 55, 57, 63
Fédération Dentaire Internationale 65
Fellowship in Dental Surgery 45
Filling teeth 23–24, 30
 materials 24, 32, 64, 66
 root canals 24
Fissure sealing 34
Fluoride 31, 65, 67
Foot treadle engine 64
Forceps, extraction 13, 14
Forceps, The 59
Fowler, J. E. H. 58
Fox, Joseph 25, 57, 63

Garretson, J. E. 58
General Dental Council 45, 65, 66
General dental services 51–53
Gold
 crowns 21, 32
 denture bases 19–20
 filling material 32
 inlays 32, 65
Guerini, V. 61
Gum (periodontal) disease 1, 3–4, 13,
 34
 prevention 31
Gum lancet 13
Gutta-percha 20, 24, 67

Harris, C. A. 58
Hesketh, Birch 38
Higher Dental Diplomas 45
Hobbs, Lucy B. 64
Hoffmann-Axthelm, W. 61
Hunter, John 57, 63
Hygienists, dental 45–46, 65

Impressions 15, 20–21
Inclined plane 25, 68
Index to dental literature 60
Instruments 13–14, 24, 25
Ivory denture bases 11–12, 15, 16, 20,
 48

*Journal of the American Dental
 Association* 60
Journals 59–60

Key 13, 14
Kingsley, N. W. 58
Koller, Carl 33, 65

Labial arch 25, 68
Leber 64
Levers 13
Libellus de dentibus 55, 63
Licence in Dental Surgery (LDS) 42,
 43, 64
Lignocaine 34
Lindsay, Lilian 57, 65
Lindsay Society 61
Literature, dental
 early sources 55
 eighteenth/nineteenth century 55,
 57–58
 history of dentistry 60–61
 journals 59–60
 reference books 58–59
Local anaesthesia 33–34
London Dental Dispensary 48
London Dental Hospital 45
London Institute for the Diseases of the
 Teeth 48, 64
London School of Dental Surgery 64

Manufacturers, dental 30
Maternity services 51
Mercury treatments 4, 22
Metropolitan School of Dental
 Science 42, 64
Microbiology 2–3, 4
Middleton, Thomas 36

Miller, W. D. 3, 58, 65
Monthly Review of Dental Surgery 59
Morrison, James Beall 64
Morton, William Thomas 28, 64

National Dental Hospital 42
National dental survey 66
National Health Service 52–53, 65
National Insurance Act (1911) 51
Nitrous oxide 27, 30, 64

Obturators 22–23, 68
Odontological Society 41, 42, 59, 64, 65
Operator for the teeth, The 55, 56, 63
Operators for the teeth 35–37
Orthodontics 24–27, 34, 58, 68

Pelicans 13
Pepys, Mrs Samuel 36
Plaque 2, 3, 31, 68
Porcelain
  crowns 22
  dentures 16–18, 63
Prevention 4, 5–7, 31
Priestley, Joseph 27
Procaine 33, 65
Punch 13

Quarterly Journal of Dental Science 59

Radiography 34, 65
Rae, James 63
Reference books 58–59
Reimplantation 13
Ring, M. E. 61
Robinson, James 28, 29, 41, 59
Rome, classical 6–7, 9, 11, 63
Root canal filling 24
Rottenstein 64
Royal Colleges 42, 43, 45, 64
Royal Dental Hospital 42
Rubber band 25
Rymer, Samuel Lee 41

St Apollonia 7
School dentistry 49–50
School Dentists Society 49
Schools, dental 43, 45, 64
Scurvy 4
Shell crowns 21, 69

Silicate cement 64
Siwak 5
Skinner, R. C. 57
Snell, James 64
Specialist journals 60
Stent's compound 21
Streptococcus mutans 4
Sugar 2, 31, 47
Suppliers, dental 30
Surgery, dental 30
Surveys, dental health 49, 52, 66
Syphilis 22

Taggart, William, H. 65
Tartar 4, 13, 69
Teething powders 47
Terro-metallic teeth 18, 19
Therapist, dental 45
Tinctures 13
Tomes, John 42, 58
Tongue scrapers 6
Toothache cures 7–9, 12–13, 47, 55, 63
Toothbrush Clubs 50, 51
Toothbrushes 5–6, 63
Tooth-coloured filling materials 32, 64, 66
Toothdrawers 35, 38
Toothpaste 7, 31, 65
Toothpicks 6
Toothpowders 6–7, 47
Toothworms 2, 9, 63
Transactions of the British Society for the Study of Orthodontics 60
Transactions of the Odontological Society of London 59
Transplantation 21
Travelling dentists 37, 38
Tube teeth 18, 19
Turbine handpiece 33, 65

University degrees, dental 45, 65

Vaccines 31
Vulcanite 64, 69
  denture bases 20, 32
  obturator 23

Wells, Horace 27, 28, 64
White, S. S. Company 60
Woodhouse, A. J. 39
Worn-down teeth 1, 21